S0-AZS-917

DISCOVER COLORS

Marie-Agnès Gaudrat and Thierry Courtin
Adapted by Judith Herbst

The title of the French edition is *À la Découverte des Couleurs.*
Adapted from the French by Judith Herbst.

First English language edition published in 1995 by
Barron's Educational Series, Inc. English translation/adaptation

Address all inquiries to:
Barron's Educational Series, Inc.
250 Wireless Boulevard
Hauppauge, New York 11788

Library of Congress Catalog Card No. 94-39683

International Standard Book No. 0-8120-6497-6

Library of Congress Cataloging-in-Publication Data
Herbst, Judith.
Discover colors / Marie-Agnès Gaudrat and Thierry Courtin ; adapted by Judith Herbst. — 1st English ed.
p. cm.
Adapted translation of: A la découverte des couleurs.
ISBN 0-8120-6497-6
1. Colors—Juvenile literature. [1. Color.] I. Courtin, Thierry. II. Gaudrat, Marie-Agnès. A la découverte des couleurs. III. Title.
QC931.4.H47 1995
535.6—dc20 94-39683
CIP
AC

Printed in France
5678 9655 987654321

THE STORY OF COLORS

Colors have a generous nature; they are there for all to see. We need to try and figure out what colors look like—they are all around us.

But just looking at colors is not the same as knowing or recognizing them. The next, more abstract step is to name them. When children give it a first try, it's always by guessing and experimenting: "Red?" "Yes, very good"..."Blue?" "No, not this time, it's green!" We have to admit that the child's guess is plausible: blue could well be called green, and green, blue. It helps to have some frame of reference. That's why, in this book, we use an animal to present each color. Getting the child to associate the color green with the crocodile, gray with the elephant, and yellow with the chick is our way of personalizing each color for easy memorization.

After a while, children will be able to understand the nuances in color—to marvel at the fact that "all blues aren't the same" or to realize that blond is related to yellow, and red to orange.

Now we are ready to enter into the curious world of mixing colors (OR into the world of alchemy) that gives us green when blue is added to yellow! The yellow duck shows us how this happens when it dives into the blue water. The mysteries of orange and purple are unfolded to us in a similar manner. And as a little finger traces the snails on their maze-like trail, the child discovers what happens to colors when they are blended together haphazardly.

Finally, the most amazing discovery of all is to see the way colors play with and against each other. Sometimes color is used to tone down, render less distinct; other times it is used to make another color more vivid. We have entrusted the zebra, the guinea hen, and the chameleon with the job of introducing these dynamics to the child.

A thank you to all the animals for lending us their unique characteristics and good humor as we help children enter the exciting world of colors.

White hen, white hen, smartly dressed,

Showing off her bright red crest.

The crocodile is very green,
From back to front and head to feet;

He keeps his white teeth sparkling clean
In case there's something good to eat.
(Yum, yum!)

This parakeet is oh so blue,
But oh my gosh! The sky is, too!
With all that blue, it seems to me
A blue bird would be hard to see.

But there are lots and lots of blues,
All sorts of blues from which to choose;
So when the blue birds come to dine,
They all wear blue and look divine.

Who's this fellow, all in yellow,
Sloshing through the rainy street?
Watch him splashing, puddle crashing,
Yellow boots upon his feet.

Thirteen yellow chickadees, singing sweetly for us;
If you know the melodies, you can join the chorus.

The elephant is big and gray,
He likes it very much that way;

But gray is nice for little mice,
In fact, they wear it everyday.

Pink, I think, is simply perfect, oh so fine for painting walls;

Pink for pearls and little piglets, pink for painters' overalls.

When seen about town, the mandrill is brown,
His red and blue mask is so striking;

Gorillas, however, are terribly clever,
A black suit is more to their liking.

Owl, owl sitting there,
Shouldn't you be in the air?
Swooping through the darkened skies,
Guided by your yellow eyes.

Oh my! I simply have to ask:
What color is the panda's mask?

Name the game this mole just loves,
And name the color of his gloves.

A bright red hat, a light brown bat,
A bulging pouch—what color's that?

Whose vest is brown? Whose beak is green?
Whose head is the reddest that you've ever seen?

What color is the penguin's coat?
The pigeon's fancy crest and throat?

Where, oh where can I find red?
Somewhere on the robin's head?
Is there red above his beak?
Maybe red runs down his cheek.

Where is red on this baboon?
Surely I will find it soon!
Maybe you should look for me;
Point to all the red you see.

Blond is really yellow but a little bit more mellow;
Though it's nice for a canary,
All the lions think it's scary.

Squirrels look swell in a deep shade of red,
Dressed in maroon from their toes to their head,
One little white spot adds just enough spice;
Golly, their outfit is really quite nice.

The bear's brown fur is dark and deep,
It keeps him warm while he's asleep.

The ermine's fur is lighter brown,
But then when winter comes to town…

And temperatures are 10 below,
Her coat turns white to match the snow.

But oh, that crow! He's black as coal,
He's blacker than the blackest hole;
Through summer, winter, spring, and fall,
He doesn't change his clothes at all.

The duckling is yellow, the water is blue
And looks so refreshingly wet;
So what can this fellow be planning to do?
He'll jump in the water, I'll bet!

And lo and behold! Well, isn't this keen!
When yellow meets blue, they turn into green!

But wait! I can show you some more color tricks;
Just look at what happens when red and blue mix.

And you can make purple as easy as pie,
So go get your paint set and give it a try.

Slugs may look a little gushy, slimy, mushy, not so sweet;
They leave yellow trails behind them
As they crawl along the street.

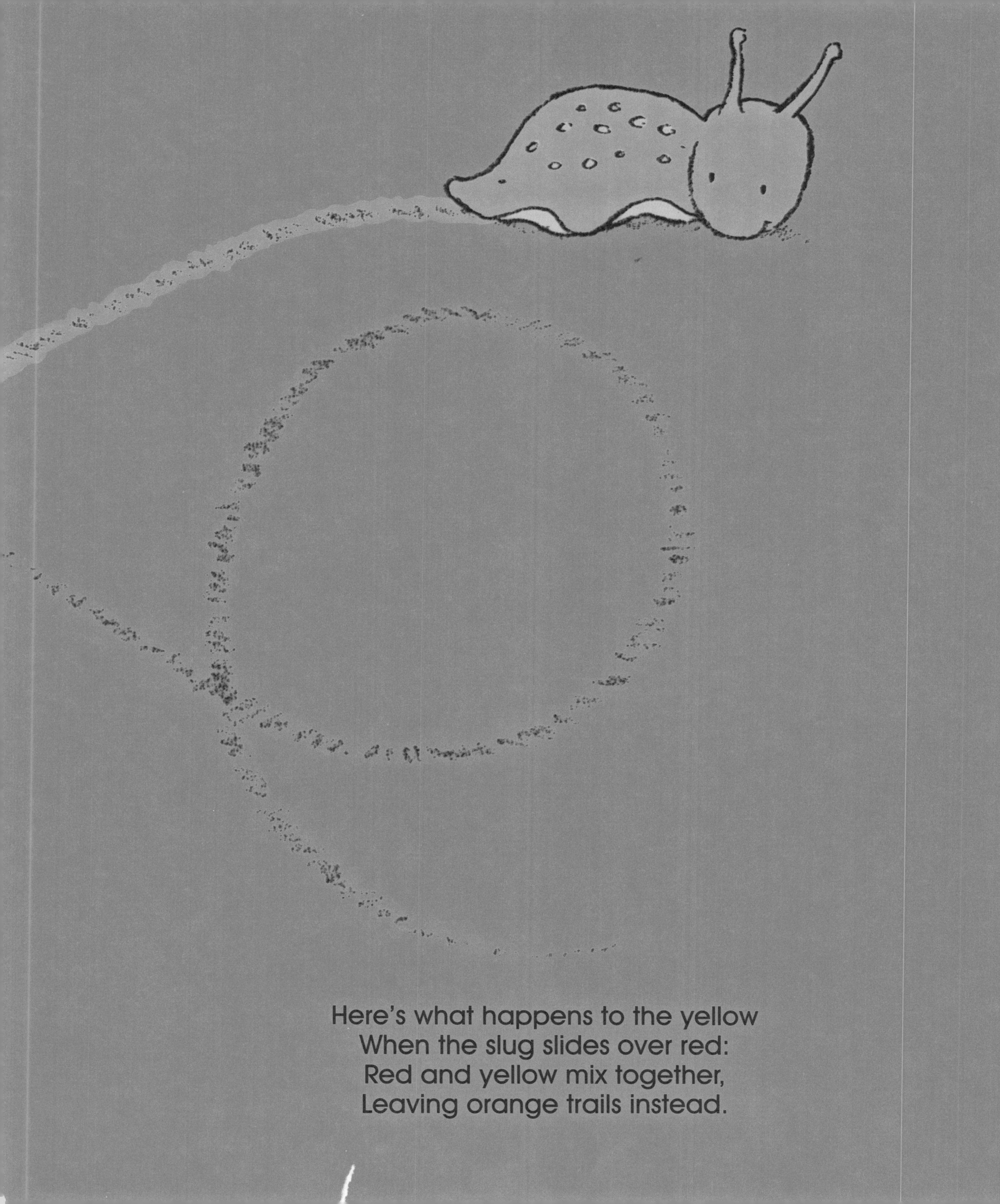

Here's what happens to the yellow
When the slug slides over red:
Red and yellow mix together,
Leaving orange trails instead.

We'll start with some orange and stir in some green
To mix up a color you might not have seen;

It won't be quite orange or green anymore,
So that's what the color called khaki is for.

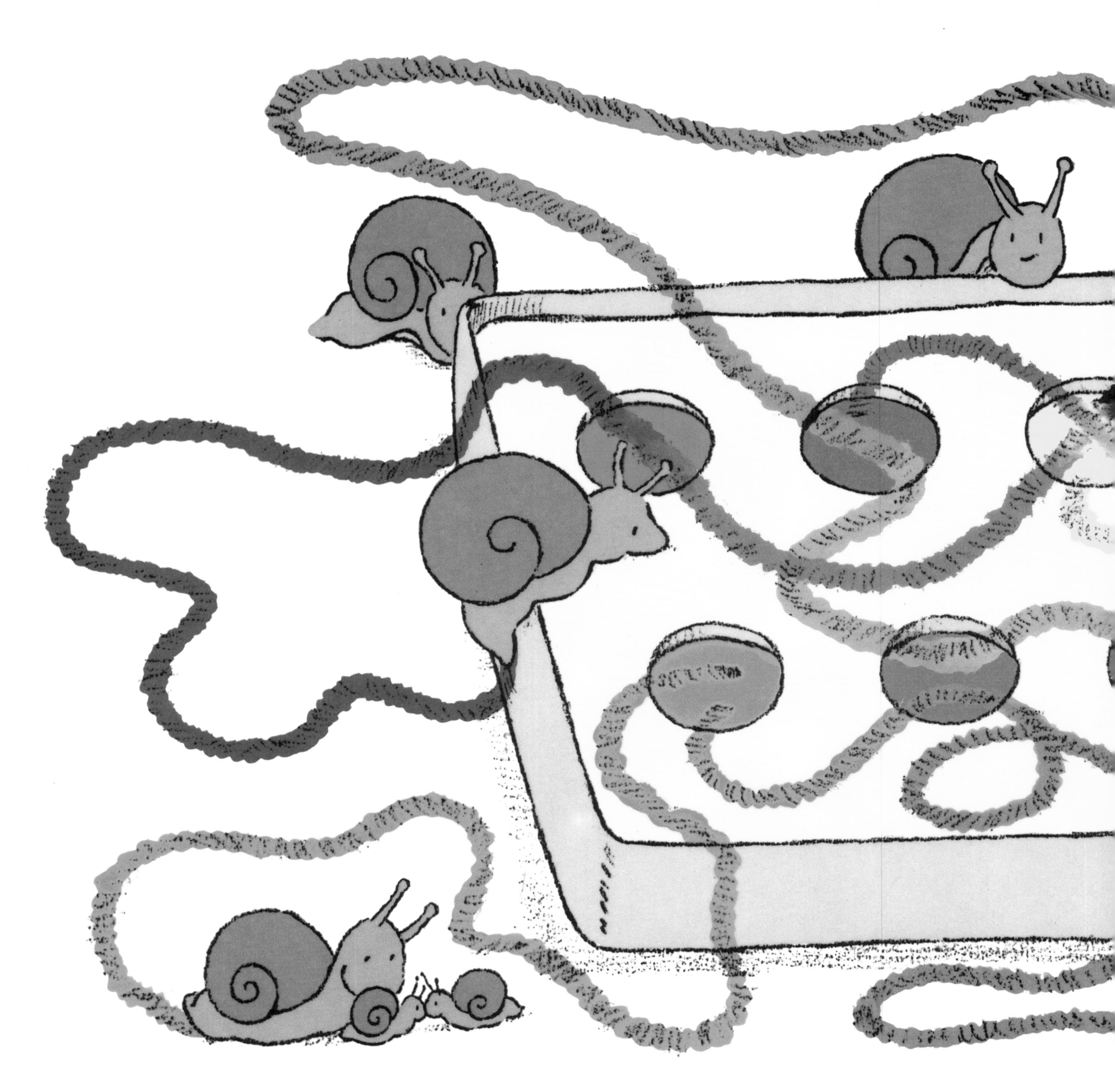

Oh, my goodness! Color trails, left behind by painting snails.

Mix the colors in your set…muddy green is what you get.

"Hey! Watch where you're jumping! There's brown on my blue,
My yellow, my green, and my red feathers, too!"

"Hey! Watch where you're standing," the hippo replied.
"It's hard to be neat when your body's this wide."

Zebra, oh zebra, you look very cute,
Proudly displaying your black and white suit;

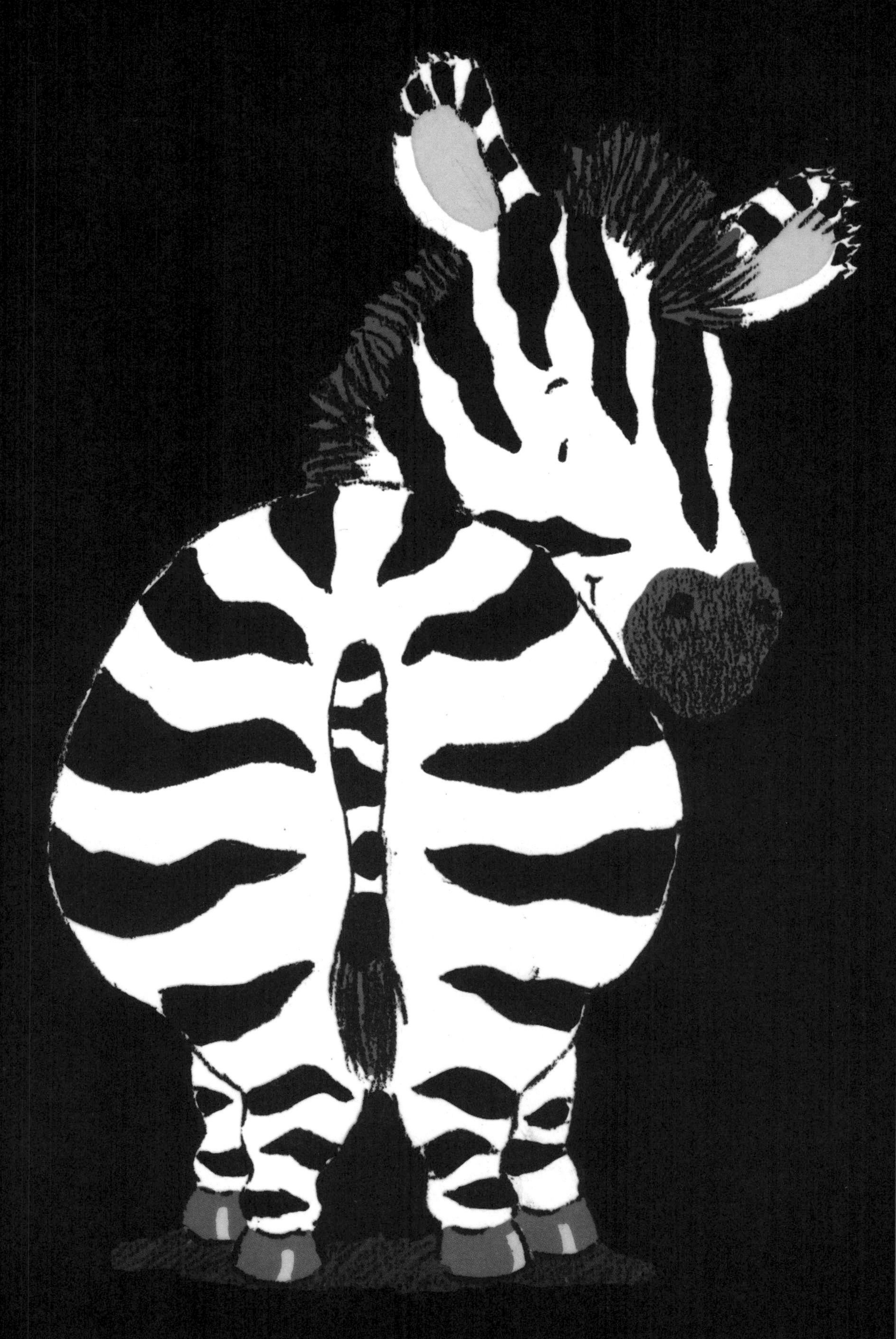

Which is there more of: the white or the black?
That all depends if you're front or you're back.

How lovely for the zebra to be striped from side to side,

How splendid for this milk cow to have spots upon her hide,

How absolutely fabulous to know this guinea hen
Has dots, and dots, and dots, and dots, and lots of dots again!

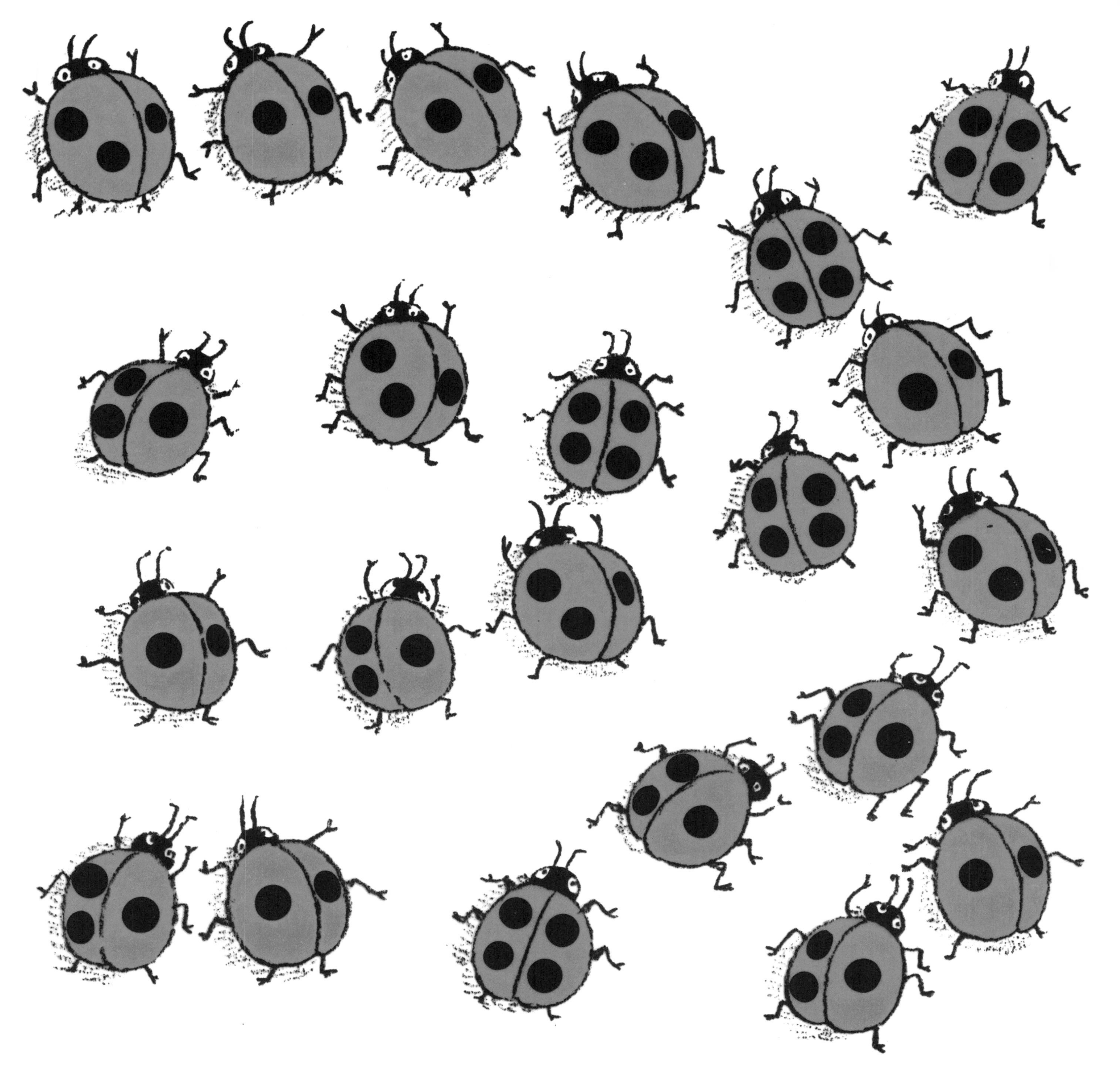

“And what a joy it is for us,” the ladybugs all said,
“To dance the summer days away decked out in black and red.”

A chameleon is glad
To dress up in plaid, he matches wherever he sits;

Be it blue plaid with red,
Or purple instead,
He's sure that his suit always fits.

The toucan's beak is quite a prize,
Despite its quite enormous size;

And like a rainbow in the sun,
It shows the colors,
EVERY ONE!

My goodness! What's this?
I just heard him hiss.
I'll bet he's the first of his kind:

A colorful fellow,
All blue, green, and yellow,
A picture I drew in my mind.

Photogravure : WESPIN
Imprimé par AUBIN IMPRIMEUR - Relié par BRUN